7.8 magnitude Earthquake in Nepal: Walking of Disaster on Himalaya

Dr. Hemant Pathak

DEDICATION

Dedicated to Shri Sainath Maharaj the all omnipotent of world the most

merciful

Also

Tribute to people of Nepal that fight from quake bravely.

CONTENTS

Foreword

In the last 10 years, several extreme events have rocked the Earth, and this is the latest calamity in that series. Nepal has witnessed some of the worst episodes in its long history both in terms of human suffering and material destruction. 7.8 magnitude Earthquake in Nepal: Walking of Disaster on Himalaya; provides a unique insight into the problems our planet faces in terms of Safe environment, and what to do about it.

This is the only books expressed comprehensive and interdisciplinary focus on, tremor crisis with the multidimensional approach.

This book made of 10 years consistently research on environmental issues, makes it ideal source for students, teachers, industrialist, environmental experts and economists.

This book provides an essential guide to researchers, it offers: various causes of increasing climate change, challenges, effect and experiences in present scenario.

Simply explained, 7.8 magnitude Earthquake in Nepal: Walking of Disaster on Himalaya; is an important book to aware common people how to plan and manage during earthquake. This is the real story of the quake tragedy that are hitting Nepal. We can learn the real lesson behind it can we find ways of dealing with it in the future.

Dr. Hemant Pathak

M.Sc. (Gold medalist), Ph. D.

Assistant Professor of Engineering Chemistry

Indira Gandhi Govt. Engineering College,

Sagar, MP, India

Glossary

acceleration The time rate of change of *velocity* of a reference point during an earthquake.

accelerometer An instrument used to measure *acceleration*. Used to measure the response of the ground or a structure to shaking in an earthquake.

amplification (seismic) The increase in surface *ground motion* at certain frequencies in unconsolidated sediments relative to the motion in solid rock.

body wave Seismic wave propagated in the interior of the earth. P and S waves are examples.

core The central part of the earth, beginning at a depth of about 2900 km, probably consisting of iron-nickel alloy; it is divisible into an outer core that may be liquid and an inner core about 1300 km in radius that may be solid.

Climate change The practice of identifying and evaluating, in monetary and/or nonmonetary terms, the effects of climate change on natural and human systems.

crust (of the earth) The outermost major layer of the earth; in Utah, ranging from 35 to 45 km thick and with a compressional seismic wave velocity (in rock) between 3.0 and 7.5 km/s.

Disaster Severe alterations in the normal functioning of a community or a society due to hazardous physical events interacting with vulnerable social conditions, leading to widespread adverse human, material, economic, or environmental effects that require immediate emergency response to satisfy critical human needs and that may require external support for recovery.

Disaster management Social processes for designing, implementing, and evaluating strategies, policies, and measures that promote and improve disaster preparedness, response, and recovery practices at different organizational and societal levels.

Early warning system The set of capacities needed to generate and disseminate timely and meaningful warning information to enable individuals, communities, and organizations threatened by a hazard to prepare to act promptly and appropriately to reduce the

	possibility of harm or loss.
earthquake	The shaking or vibrating of the ground caused by the sudden release of energy stored in rock beneath the earth's surface.
earthquake hazard	Any physical phenomenon associated with an earthquake that may produce adverse effects on human activities.
earthquake source	The origination point of earthquake energy release.
epicenter	The point on the surface of the earth directly above the point where the first rupture and first earthquake motion occur.
fault	A fracture in the earth along which the two sides have been displaced relative to each other.
Global Warming	The infrared radiative effect of all infrared-absorbing constituents in the atmosphere. Greenhouse gases, clouds, and (to a small extent) aerosols absorb terrestrial radiation emitted by the Earth's surface and elsewhere in the atmosphere. These substances emit infrared radiation in all directions, but, everything else being equal, the net amount emitted to space is normally less than would have been emitted in the absence of these absorbers because of the decline of temperature with altitude in the troposphere and the consequent weakening of emission.
geomorphology	The study of the origin and character of landforms.
hypocenter	The point within the earth where earthquake rupture begins; the focus of an earthquake.
intensity (earthquake)	A subjective numerical index describing the severity of ground shaking in an earthquake in terms of the effect on objects and humans.
landslide	The perceptible downward sliding or falling of masses of rock or soil; can include

earthflows, debris flows, rock avalanches, and rock falls.

Love wave	A type of seismic surface having only horizontal motion transverse to the direction of propagation.
magnitude (earthquake)	A number that characterizes the size of an earthquake by measuring the motions recorded by a seismograph and correcting for the distance to the epicenter of the earthquake.
Mitigation	Actions taken to avoid, reduce, or compensate for the effects of environmental damage. Among the broad spectrum of possible actions are those that restore, enhance, create, or replace damaged ecosystems.
P wave	A seismic wave that involves particle motion in the direction of propagation. It is the fastest traveling wave generated by an earthquake and therefore the first to arrive at any point.
risk evaluation, risk reduction, and risk management	All relate to exposure of life and property to earthquake hazards.
rock avalanche	A large mass of rock, sliding or flowing very rapidly under the force of gravity.
seismic wave	An elastic wave generated in the earth by an earthquake or explosion.
seismology	The study of earthquakes, earthquake sources, and the propagation of seismic waves.
seismometer	The sensor that detects the seismic wave energy and transform it into an electric voltage.
tectonics	branch of geology dealing with structure and deformation of the earth's crust.

1. Introduction

Earthquakes are a great threat to environmental stability and life in the Himalayan region as almost the entire region is prone to high seismic activity. The region has been hit by earthquakes of varying intensities in the past and similar threats remain imminent.

The latest event in that endless process came just before midday local time on the 25 April 2015, when the section that holds up India slipped under the Eurasian plate.

Nepal is the victim of most deadly earthquake in past 81 years, the worst disaster in history on the world's highest peak. The 7.8-magnitude quake hit Nepal on Saturday destroying buildings in Kathmandu and severely affecting rural areas across the region. The effects were immediate and horrific, buildings collapsed over the region, leaving nearly 10,000 dead and many more injured.

Eight million people have been affected by the massive earthquake in Nepal, more than a quarter of the country's population. International aid has started arriving but there is still huge need - 1.4 million require food aid, the UN said.

The Himalayas were formed by a head on collision of the Indian and Eurasian plates, and the Indian plate continues to push the Asian plate northward at the rate of about 2 cm per year. This means that in every 100 years India moves 200 cm north against the Asian plate, and this colliding force builds up pressure continually for several years and is released in the form of earthquakes.

The death toll has risen to more than 10000, with almost 18,000 injured, officials say. Nepal and surrounding areas have continued to experience aftershocks. Thousands in Kathmandu, Nepal's capital, spent a third night outside, too afraid to go back into their houses.

Fig.1

According to initial estimations and based on the latest earthquake intensity mapping, eight million people in 39 districts have been affected, of which over two million people live in the 11 severely affected districts.

The situation is critical in the remote rural regions towards the epicenter of earth quake. 90% of houses flattened in Gorkha district close to the epicentre. Many have also lost livestock and have little food. Water, food and electricity are in short supply and there are fears of outbreaks of disease.

Villages around the epicentre were very difficult to reach - cut off by landslides - and that bad weather was hampering helicopter access.

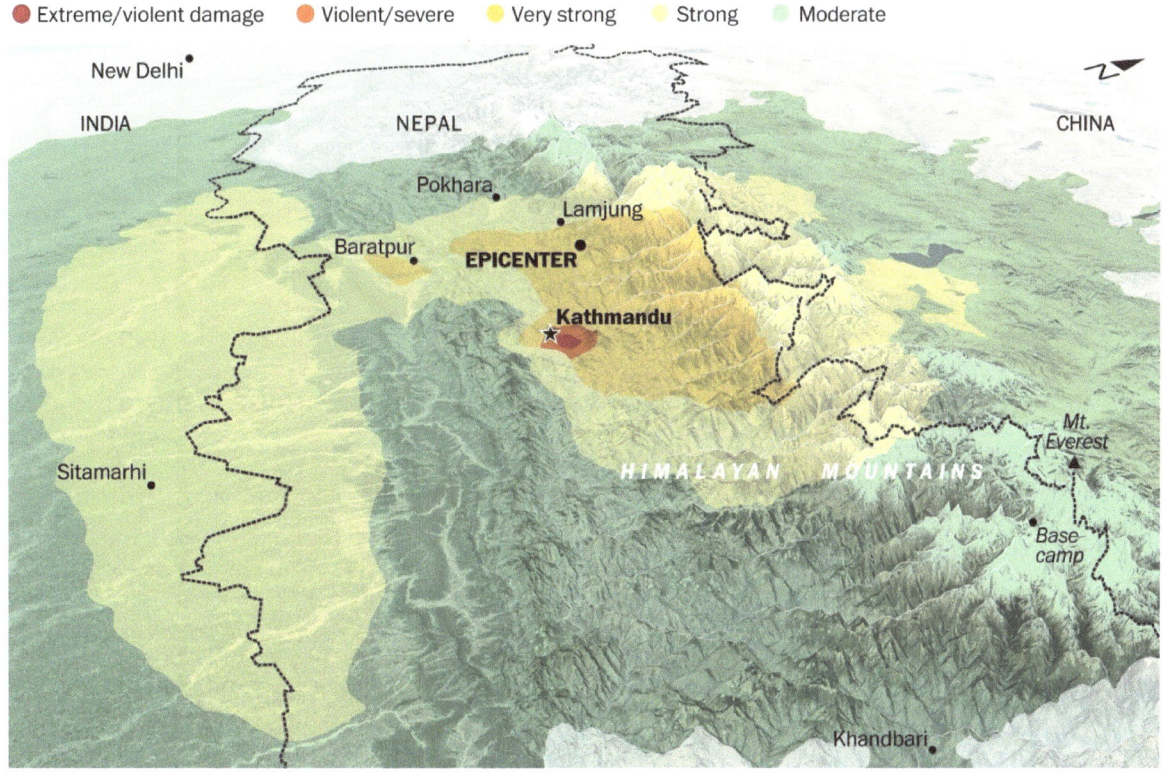

Fig.2

The Nepal government has pleaded for overseas aid - everything from blankets and helicopters to doctors and drivers. also urge foreign countries to give us special relief materials and medical teams. Almost the entire Nepali army and police have joined the search and rescue operations. Many countries have sent aid including India, China, the UK and US.

Hospitals are unable to cope with the huge numbers of people in need of medical attention and some Nepalis have complained of aid being slow to reach them.

1.4 million people are in need of food assistance. Of these, 750,000 people live near the epicenter in poor quality housing. Impact on agriculture based livelihoods and food security is expected to be extremely high. Immediate needs for health include medical tents, medication, surgical kits and body bags.

But outside the capital, many of the worst-hit villages in the ridges around Katmandu remain a black hole, surrounded by landslides that make them inaccessible even to the country's armed forces.

The Nepali authorities began airdropping packages of tarpaulins, dry food and medicine into mountain villages, but an attempt to land helicopters was abandoned.

Hinduism's holiest sites, the Pashupati temple in the Katmandu Valley, and Lumbini, a pilgrimage site in the southern plains believed to be the Buddha's birthplace, had been spared.

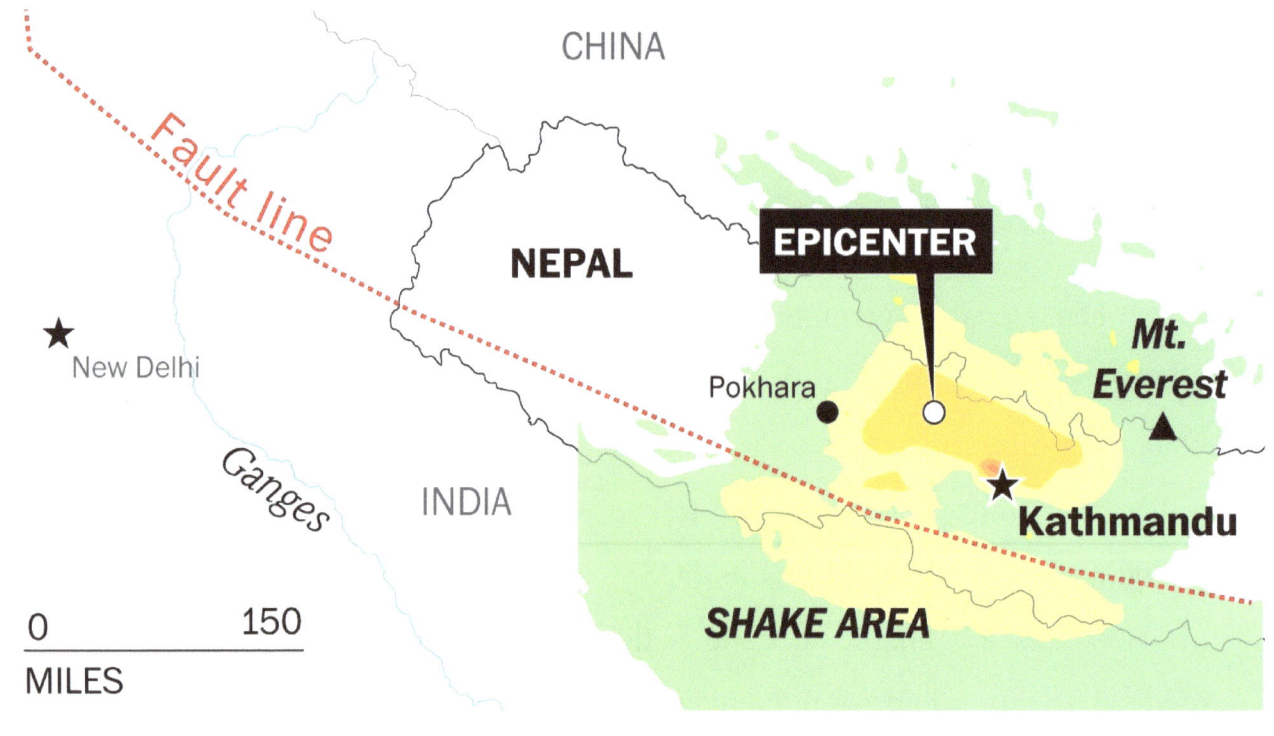

Fig.3

2. **What is Earthquakes?**

Most destructive of natural hazards. Earthquake occurs due to sudden transient motion of the ground as a result of release of elastic energy in a matter of few seconds.

The impact of the event is most traumatic because it affects large area, occurs all on a sudden and unpredictable. They can cause large scale loss of life and property and disrupts essential services such as water supply, sewerage systems, communication and power, transport etc. They not only destroy villages, towns and cities but the aftermath leads to destabilize the economic and social structure of the nation.

Word earthquake is used to describe any seismic event, whether natural or caused by humans that generates seismic waves. Earthquakes are caused mostly by rupture of geological faults, but also by other events such as volcanic activity, landslides, mine blasts, and nuclear tests. An earthquake's point of initial rupture is called its focus or hypocenter. The epicenter is the point at ground level directly above the hypocenter.

3. History of quake in Himalaya

Plate tectonics studies reveal that the Himalayan mountain ranges were formed when Indo-Australian plate collided with the Eurasian plate. The Indian subcontinent, once part of the supercontinent called Gondwanaland, which consisted also of present-day Africa and Antartica, broke away about 100 million years ago and crawled northwards across the Tethys Sea before ramming into Asia.

Tectonics of the Himalayas. Around 45ma, during the Eocene, India collided with Eurasia; this was the end result of India's northward migration caused by the subduction of the Tethyan Ocean beneath the Eurasian Plate. The Indian plate slid under the Asian landmass. Its upper layers peeled and thrust upward, forming the Himalayan ranges.

Four great earthquakes of Himalaya i.e., Assam earthquake of 1897, Kangra earthquake of 1905, Nepal-Bihar earthquake of 1934 and Assam earthquake of 1950 rocked the Himalayan Kingdom of Nepal whose magnitude exceeds 8.0 Richter scale.

The earthquake of 1833 with magnitude 7.8 occurring at a distance of 50kms NE of Kathmandu also affected the Kathmandu Valley. In the last few decades, the major amongst them are the Kinnaur earthquake of 1975, Dharchula earthquake of 1980 and the Uttarkashi earthquake of 1991, which resulted in tremendous loss of life and property.

4. Causes of Earthquakes

An Earthquake (also known as a quake, tremor or temblor) is a series of underground shock waves and movements on the earth's surface caused by natural processes within the earth's crust.

It is estimated that around 500,000 earthquakes occur each year, detectable with current instrumentation. About 100,000 of these can be felt. Minor earthquakes occur nearly constantly around the world.

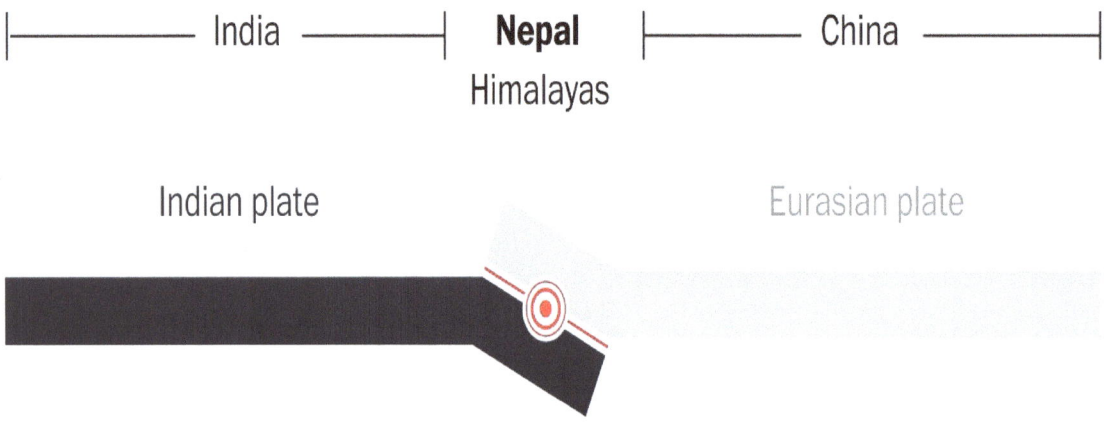

Fig.4

The seismicity, seismism or seismic activity of an area refers to the frequency, type and size of earthquakes experienced over a period of time.

Millions of years, the squeezing has crushed the Himalayas like a concertina, raising mountains to heights of several miles and triggering earthquakes on a regular basis from Pakistan to Burma. Nepal's quake was neither unusual nor unexpected, although it was larger than most.

In the 81 years since the 1934 Bihar earthquake, the land mass of India has been pushed about 12 feet into Nepal. Think of all that movement getting stored in a giant spring lying under Nepal. The spring is stuck on a broad, rough surface which we call a fault plane (a fault line is what we see when it emerges from the ground).

Sometimes, energy stored in the spring gets big enough to slip catastrophically, releasing all that pent-up strain and generating shaking strong enough to destroy buildings and kill people over a huge area. The bigger the area that slips, and the larger the pent-up energy, the greater the damage.

This slip took place over an area about 1,000 to 2,000 square miles over a zone spanning the cities of Kathmandu and Pokhara in one direction, and almost the entire Himalaya mountain width in the other. A part of India slid about one to 10 feet northwards and underneath Nepal in a matter of seconds.

Giant cracks (fault zones) formed along its northern perimeter. Since then India has penetrated some 2000km into Asia leading to the creation of the Himalayan mountain range. Recent GPS measurements put the present convergence rate between India and Asia at 58 ± 4mm.yr-1 [Bilham et al., 1997], roughly a third of this motion (17.52 ± 2mm.yr-1) is accommodated by the Himalayan thrust system meaning that the Kingdom of Nepal is shortening by about 2cm.yr-1.

The seismic zoning map of India shows that the entire Himalayan region lies in Zone IV and V, which correspond to MMI of VIII and >IX respectively. Seismic studies show that great earthquakes (M>8) tend to recur in cycles of 200 – 300 years along the length of the Himalayas.

At the Earth's surface, earthquakes manifest themselves by shaking and sometimes displacement of the ground. When the epicenter of a large earthquake is located offshore, the sea bed may be displaced sufficiently to cause a tsunami. Earthquakes can also trigger landslides, and occasionally volcanic activity.

5. Effects of Earthquakes on Nepal

Shaking and ground rupture are the main effects created by earthquakes, principally resulting in more or less severe damage to buildings and other rigid structures. The severity of the local effects depends on the complex combination of the earthquake magnitude, the distance from the epicenter, and the local geological and geomorphological conditions, which may amplify or reduce wave propagation. The ground-shaking is measured by ground acceleration.

About 1.45 million people live in Kathmandu, the majority in poorly constructed homes not designed to withstand the kind of shaking. Nepal has a per capita income of around $1,350 among the lowest in the world.
Meeting building codes in new construction, or taking on expensive retrofitting, is way beyond the means of most. To make matters worse, the valley itself appears to focus the destructive shaking of earthquake waves.

A number of buildings collapsed in the center of the capital, the ancient Old Kathmandu, including centuries, old temples and towers. Among them was the Dharahara Tower, one of Katmandu's landmarks built by Nepal's royal rulers in the

1800s and a UNESCO-recognized historical monument. It was reduced to rubble and there were reports of people trapped underneath. At least 28 were killed and dozens more were injured on Mount Everest, where the quake launched an avalanche. Dozens if not hundreds remain trapped under mounds of rubble. The nation's capital of Katmandu was particularly hard hit. The world is in the middle of a 15-year intense seismic activity period that could last until 2020, leaving open the strong possibility of a strong earthquake occurring somewhere on the planet, possibly in the Central Himalayas.

Specific local geological, geomorphological, and geostructural features can induce high levels of shaking on the ground surface even from low-intensity earthquakes. This effect is called site or local amplification. It is principally due to the transfer of the seismic motion from hard deep soils to soft superficial soils and to effects of seismic energy focalization owing to typical geometrical setting of the deposits.

Ground rupture is a visible breaking and displacement of the Earth's surface along the trace of the fault, which may be of the order of several metres in the case of major earthquakes. Ground rupture is a major risk for large engineering structures such as dams, bridges and nuclear power stations and requires careful mapping of existing faults to identify any which are likely to break the ground surface within the life of the structure.

6. Effects of Earthquakes on Other Countries

- **INDIA**

Indian officials confirmed more than 60 fatalities in the states of Uttar Pradesh, Bihar and West Bengal. Assam and Bihar states and the Darjeeling hill region reported many collapsed buildings.

- **BANGLADESH**

High-rise buildings cracked and swayed in several Bangladeshi cities, including the capital, Dhaka. At least four quake-related deaths have been reported.

- **CHINA**

China's state media reported at least 12 deaths in the Tibet region.

7. Prediction and Measurement of an earthquake

There is no way to predict an earthquake or its intensity, just sampling seismic data over the past 150 years points to the possibility of a huge quake shaking up the area over the next few years. Despite considerable research efforts by seismologists, scientifically reproducible predictions cannot yet be made to a specific day or month. However, for well-understood faults the probability that a segment may rupture during the next few decades can be estimated.

There are periods of intense seismic activity with high-magnitude earthquakes for near 15 years followed by a inter-seismic activity over next 30-40 years.

There are indicators that considerable stress is building up in certain segments of the Himalayan range. This high strain is locked in parts of Himalayas, ranging from near Himachal Pradesh towards the west of Nepal. However, even plate movement and convergence is a very slow process.

Earthquakes are measured using observations from seismometers. The moment magnitude is the most common scale on which earthquakes larger than approximately 5 are reported for the entire globe. The more numerous earthquakes smaller than magnitude 5 reported by national seismological observatories are measured mostly on the local magnitude scale, also referred to as the Richter magnitude scale. These two scales are numerically similar over their range of validity. Magnitude 3 or lower earthquakes are mostly almost imperceptible or weak and magnitude 7 and over potentially cause serious damage over larger areas, depending on their depth. The largest earthquakes in historic times have been of magnitude slightly over 9, although there is no limit to the possible magnitude.

Earthquakes can be recorded by seismometers up to great distances, because seismic waves travel through the whole Earth's interior. The absolute magnitude of a quake is conventionally reported by numbers on the moment magnitude scale (formerly Richter scale, magnitude 7 causing serious damage over large areas), whereas the felt magnitude is reported using the modified Mercalli intensity scale (intensity II–XII).

Every tremor produces different types of seismic waves, which travel through rock with different velocities:

- Longitudinal P-waves (shock- or pressure waves)
- Transverse S-waves (both body waves)
- Surface waves — (Rayleigh and Love waves)

Propagation velocity of the seismic waves ranges from approx. 3 km/s up to 13 km/s, depending on the density and elasticity of the medium. In the Earth's interior the shock- or P waves travel much faster than the S waves (approx. relation

1.7 : 1). The differences in travel time from the epicentre to the observatory are a measure of the distance and can be used to image both sources of quakes and structures within the Earth. Also the depth of the hypocenter can be computed roughly.

In solid rock P-waves travel at about 6 to 7 km per second; the velocity increases within the deep mantle to ~13 km/s. The velocity of S-waves ranges from 2–3 km/s in light sediments and 4–5 km/s in the Earth's crust up to 7 km/s in the deep mantle. As a consequence, the first waves of a distant earthquake arrive at an observatory via the Earth's mantle.

On average, the kilometer distance to the earthquake is the number of seconds between the P and S wave times . Slight deviations are caused by in homogeneities of subsurface structure. By such analyses of seismograms the Earth's core was located in 1913 by Beno Gutenberg.

Earthquakes are not only categorized by their magnitude but also by the place where they occur. The world is divided into 754 Flinn–Engdahl regions (F-E regions), which are based on political and geographical boundaries as well as seismic activity. More active zones are divided into smaller F-E regions whereas less active zones belong to larger F-E regions.

Table. 1 : Richter magnitude and its effects

Richter Magnitude	Earthquake Effects
Less than 3.5	Generally not felt, but recorded.
3.4-5.4	Often felt, but rarely causes damage.

Under 6.0	At most slight damage to well-designed buildings. Can cause major damage to poorly constructed buildings over small regions.
6.1-6.9	Can be destructive in areas up to about 100 kilometers across where people live.
7.0-7.9	Major earthquake. Can cause serious damage over larger areas.
8 or greater	Great earthquake. Can cause serious damage in areas several hundred kilometers across.

8. Role of human activity to produce earthquakes

Earthquakes are caused by movement of the Earth's tectonic plates. Human activity can also produce earthquakes. Four main activities contribute to this phenomenon:

1. Storing large amounts of water behind a dam

2. Drilling and injecting liquid into wells

3. Coal mining

4. Oil drilling

Global warming have capacity to change the pressure attribute of the earth sphere, These pressure changes have impact on the pressure belts in the atmosphere have impact on the ocean water and subsequent impact on the inside area of the earth. The disappearing ice, sea-level rise and floods already forecast for the 21st century are inevitable as the earth warms and weather patterns change and they will shift the weight on the planet.

It's the ebb and flow of rainwater in the great river deltas of India and Bangladesh, and the pressure that puts on the grinding plates that make up the surface of the planet. Recently discovered, that causal factor is seen by a growing body of scientists as further proof that climate change can affect the underlying structure of the Earth.

Climate change has been causing enormous and disturbing changes in the size and shape of the South Asian monsoon, while human tampering has played a part in floods. That makes the 150 giga tonnes of extra water that collects in Bangladesh after a heavy monsoon season, tilting the Indian plate.

Also the coming and going of the weight of the monsoon rains was causing energy to build up at the edge of the plate. Due to the increased global temperature, Arctic and associated areas is melted, earth are affected due to alteration in the pressure on the earth crust. water level on the seas and oceans increase due to the increased water quantity caused by the global warming, impact on the inside zones of earth can cause many tremors.

As sea levels climb remorselessly, the load-related bending of the crust around the margins of the ocean basins might in time act to sufficiently 'unclamp' coastal faults, allowing them to move more easily; at the same time acting to squeeze magma out of susceptible volcanoes that are primed and ready to blow.

9. Aftershocks

The Himalayas bordering Nepal are the result of an endless shoving match between the Eurasian and Indian tectonic plates, a natural phenomenon which can have devastating consequences.

An aftershock is an earthquake that occurs after a previous earthquake, the mainbshock. An aftershock is in the same region of the main shock but always of a smaller magnitude.

In Nepal aftershocks were at the relatively shallow depth of about six miles below the Earth's surface. Shallow quakes cause more damage than deeper ones that have miles of earth to absorb the shaking. Here are all the earthquakes greater than 5-magnitude measured in the country since the 7.8-magnitude quake on Saturday.

BARPAK, NEPAL

Fig.5

If an aftershock is larger than the main shock, the aftershock is redesignated as the main shock and the original main shock is redesignated as a foreshock. Aftershocks are formed as the crust around the displaced fault plane adjusts to the effects of the main shock.

10. Factors contributing to vulnerability and elements of risk

Several key factors contribute to vulnerability of human populations:

1. ☐Location of settlements in seismic areas, especially on poorly consolidated

soils, on ground prone to landslides or along fault lines.

2. Building structures, such as homes, bridges, dams, which are not resistant to ground motion.

3. Unreinforced masonry buildings with heavy roofs are more vulnerable than lightweight wood framed

structures. Dense groupings of buildings with high occupancy.

4. ☐Lack of access to information about earthquake risks.

11. Conclusion and Discussion

Community preparedness is vital for mitigating earthquake impact. Most effective programs are formal and initiated at the community level with support by local or national governments.

Educating the public on the causes and characteristics of an earthquake and what they should do if one occurs. Public officials and services must make contingency plans to react to the emergency. Public awareness programs can be designed to reach every vulnerable person and may significantly reduce the social and material costs of an earthquake.

Public officials and services must make contingency plans to react to the emergency. Activities the public sector may undertake include:

1. Reviewing the structural soundness of facilities that are essential for disaster response such as hospitals, fire stations, communications installations and upgrading them.☐

2. Training teams for search and rescue operations or ensuring the rapid availability of detection equipment, disaster assessment, Identifying safe sites, trauma care

3. Planning for an alternative water supply, clear streets for emergency access,

emergency communication systems and messages to the public regarding their security

4. ☐Training teams to determine if buildings are safe for reoccupancy, ☐Preparing flood plans for susceptible areas and Coordinating preparations with voluntary organizations

12. Earthquake Response Plan and Mitigation

The emergency measures of evacuation, search, rescue and relief form important action plans in disaster management. To prepared plan whereby the receipt of a signal of an impending disaster would simultaneously energise and activate the mechanism of response and mitigation without loss of crucial time. There is no standard solution to mitigate a disaster risk. The goal is to minimize the impact of disaster. Possible risk reduction measures designed and built structures to withstand ground shaking. Develop earthquake resistant construction techniques.

13. References

1. Climate Change 1995: The Science of Climate Change. Contribution of Working Group I to the Second Assessment Report of the Intergovernmental Panel on Climate Change [Houghton, J.T., L.G. Meira Filho, B.A. Callander, N. Harris, A. Kattenberg, and K. Maskell (eds.)]. Cambridge University Press, Cambridge, UK and New York, NY, USA, 572 pp.

2. www.quake.wr.usgs.gov Earthquake Reporting Service: U.S. Geological Survey and UC Berkeley service for earthquake reporting.

3. www.nicee.org: website of The National Information Center of Earthquake Engineering (NICEE).

4. www.imd.ernet.in/section/seismo/static/welcome.htm

5. www.bmtpc.org

6. www.earthquake.usgs.gov

7. www.neic.cr.usgs.gov

8.www.asc-india.org

ABOUT THE AUTHOR

Dr. Hemant Pathak held positions as Assistant Professor in the department of chemistry, Govt. Indira Gandhi Engineering College, Sagar, MP, India. He had extensive experience in teaching, research and administrative management.

Dr. Pathak received his Ph.D. degree in chemistry from Dr. Hari Singh Gour Central University, Sagar, India and M.Sc. Gold medalist from Jiwaji University, Gwalior. He has published 38 books (including e- books) and more than 50 research papers in reputed International and National journals and received several awards. He is a member of editorial boards and reviewer boards of several international journals and societies. His area of specialization includes Engineering Chemistry, Energy audits and Environmental Pollution management.

www.ingramcontent.com/pod-product-compliance
Lightning Source LLC
Chambersburg PA
CBHW050425180526
45159CB00005B/2410